WATER WONDERFUL WORLD

All Kinds Of Water Facts And Fun Activities To Spark A Child's Love For Science, Learning And Discovery

Joseth Howell

Library of Congress Control Number: 2021913261
ISBN: 978-0-578-94448-7

Edited by Richard Bryan, richi_bryan@yahoo.com
Layout design by Jose O., joseey245@gmail.com

For more information and speaking engagements, send inquiries to josethabara@gmail.com

Also available as an ebook.

Printed in the United States of America.

DEDICATION

To my husband, Orville, and children, Samuel and Nathan, whose unwavering support made this dream of a book which you now read a reality. —JH

Table of Contents

Our World 1
Water Domain 4
Water cycle 9
Water Tides 15
Water Habitat 22
Water States 24
Water Types 31
Water Uses 34
Water Conservation 40
Water Fun Activities 47
Glossary 49
Index 51

Our World

What an amazing world we live in,
with lots of interesting things to discover,
and endless possibilities to imagine.
In Science and Math, there is so much to uncover
about our world we call Planet Earth,
the third planet from the sun.
The rolling hills and flowery meadows,
the lofty mountains and giant trees,
the dazzling sun and twinkling stars,
barely give us a glimpse of what a wonder Planet Earth is.

Amazing Fact

The summit of Mount Chimborazo is the farthest point on Earth from the Earth's center. It is located in Ecuador in South America.

Think of the silvery dust that falls
off the wings of moths as they flutter in the moonlight,
and don't forget the fireflies and grasshoppers too.
From tiny wriggly worms to enormous elephants,
colorful peacocks to giraffes so tall,
We see beauty everywhere on land and in space.
But can you think of one thing we cannot chase
found in the air and on the ground?
Yes, you guessed it! Water. Just look all around.

Amazing Fact

The earth is the only planet in the solar system that supports life.

Review Questions

1. What position is Planet Earth from the sun?
   (a) 1st (b) 3rd (c) 6th
2. Water can be found beneath the earth and in the air. *True* or *False*.
3. Look around your backyard.
   Name three things in nature you can see.
   1. ________ 2. ________ 3. ________

Amazing Fact

Water is the most abundant substance on earth. It covers almost three-quarters of the earth's surface which means the earth is mostly covered in water.

Water Domain

Amazing Fact
Raindrops are round and not tear-drop shaped.

Up in the sky above,
and down on land below,
there is water, water everywhere!
From the skies, the clouds downpour
as raindrops.
Between the rocks, water flows
out into rivers, lakes, seas and
oceans.

Amazing Fact

A rain cloud weighs about 100,000 tons which is equivalent to the weight of 14,000 giant adult elephants.

Beneath us the earth erupts, spewing water high enough into the air to form cascading waterfalls and sparkling springs to the delight of us all. From the North Pole, all the way to the South Pole, across hot desserts and regions cold, water makes up 71 percent of our world. Water exists all around us as snow-white glaciers basking in the sun, and in creeks, rivers, lakes, seas and oceans.

Amazing Fact

Approximately 1 million miles of pipelines carry water across the United States and Canada. Did you know that's enough pipeline to circle the earth 40 times?

In reality, these bodies of water are just one large body of water separated by big chunks of land we call continents.

The Pacific, Atlantic, Indian, Arctic and Southern Oceans are the largest bodies of water which stand as monuments.

The largest of them is the Pacific Ocean

while the smallest of them is the Arctic Ocean.

If all the water in the oceans combined could be emptied out,

we would be able to fit three times all of the land on earth in it.

There are 165 major rivers in the world.
The Roe River is the shortest of them all.
Located in the United States at 201 feet,
it's just long enough to circle one big mall.
The longest river is the Nile,
found on the continent of Africa.
It flows through several countries there,
and measures 4,258 miles in length.

Amazing Fact
Ice floats on water because it is less dense than water.

Review Questions

1. In which country is the Roe River found?
   (a) United States (b) Brazil (c) China
2. Which of these rivers is the longest in the world?
   (a) Roe River (b) Amazon River (c) Nile River
3. Identify the smallest ocean in the world.
   (a) Indian Ocean (b) Pacific Ocean (c) Artic Ocean

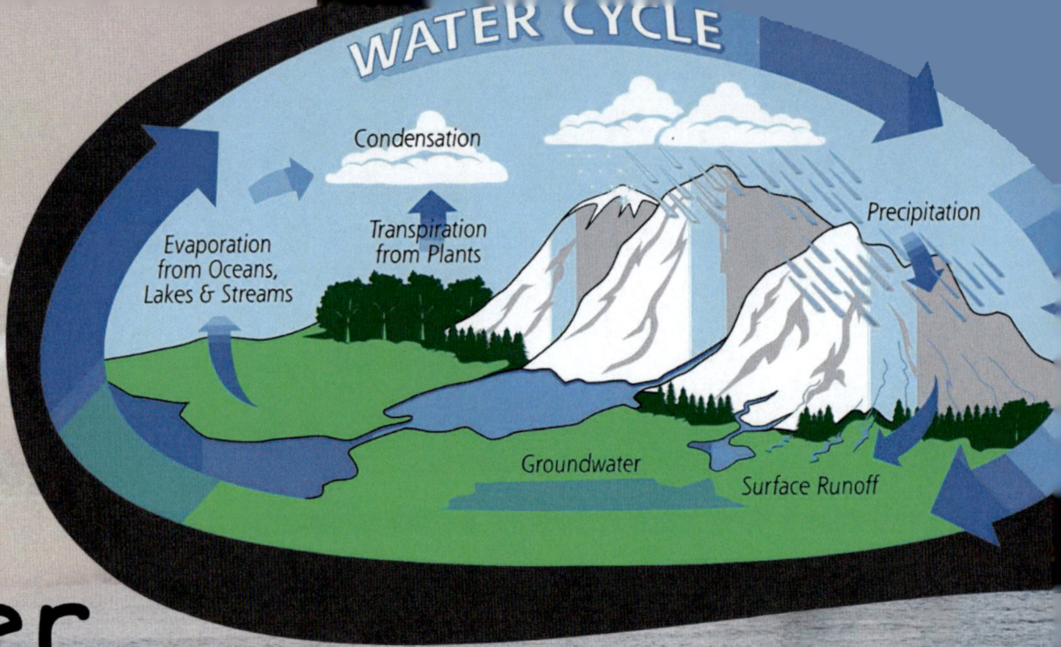

Water Cycle

As the sun heats up the land and sea,
the water on the surface changes from liquid to gas.
We refer to this process as evaporation.
And the gas is called water vapor.

As water vapor rises through the atmosphere,
it comes in contact with cool air.
The cool air causes the vapor to turn into water droplets.
We refer to this process as condensation.

As more water droplets form,
they bind together to form clouds.
When the clouds get too heavy to bear up their weight,
they fall as rain, whether early or late.

Amazing Fact

The largest glacier in continental Europe is Jostadalsbreen found in the country of Norway. It covers an area of 487 square kilometers (183 sq. miles) and has an ice thickness of about 600 meters (2,000 ft).

Clouds also let down sleet, and snow, sometimes making it difficult for us to move around. But it's all part of the process of precipitation, where clouds provide water for underground reservoirs, seas and oceans.

Plants and animals are not left out here,
for they too release water vapor into the atmosphere.
We refer to this process as transpiration.
And together these processes sum up the water cycle.

Amazing Fact

Did you know that Australia receives an average rainfall of 165mm (6.5 inches) annually, making it the driest inhabited continent in the world?

Water you see is always on the move,
following a path through different states
from the clouds to land and sea
across green hills, glaciers, rivers and lakes.

Amazing Fact

It is estimated that 110 million gallons of water are drunk every day in the United States.

Review Questions

1. The process by which water changes from liquid to gas is known as ________________.

   (a) Precipitation (b) Evaporation (c) Aspiration

2. How do plants and animals contribute to keeping the water cycle going?

3. How would you explain the different processes that occur in the water cycle to a friend?

Water Tides

In ebbs and flows the tides come and go
as seas rise and fall to shore.
This they do, all day, everyday
never resting to take a break.
But just what are tides, one may ask?
I'm glad you asked I must say...

Amazing Fact

At the Bay of Fundy in Canada located in North America, tides can reach a peak of up to 16.3 meters (53.5 feet), the highest in the world."

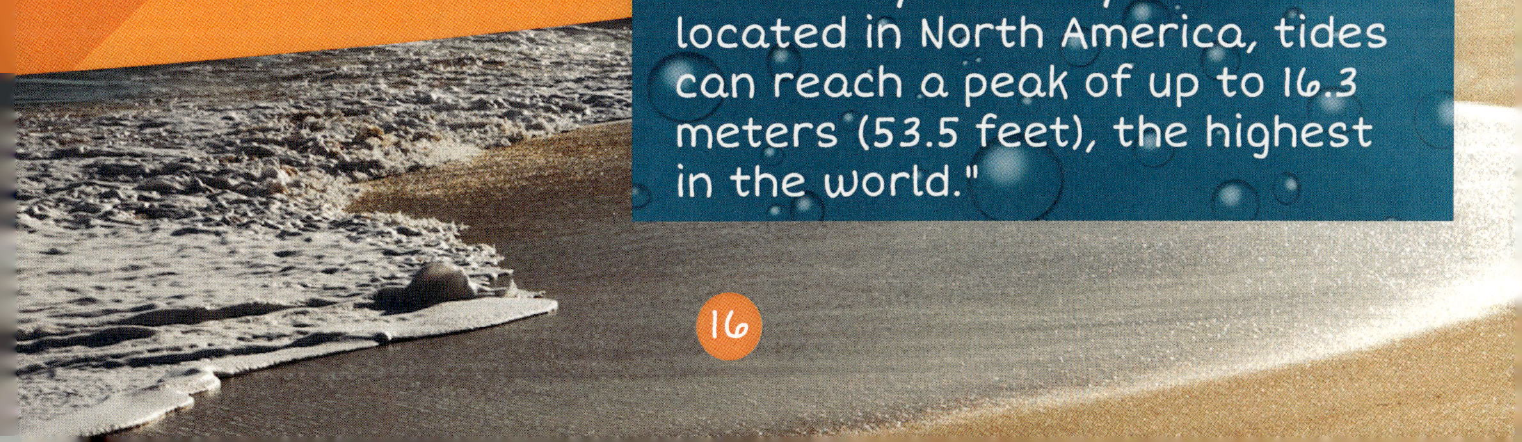

Amazing Fact

A milking dairy cow drinks about 30 to 50 gallons of water in a day. That is equivalent to 480 to 800 cups of water. Can you drink that much water in a day?

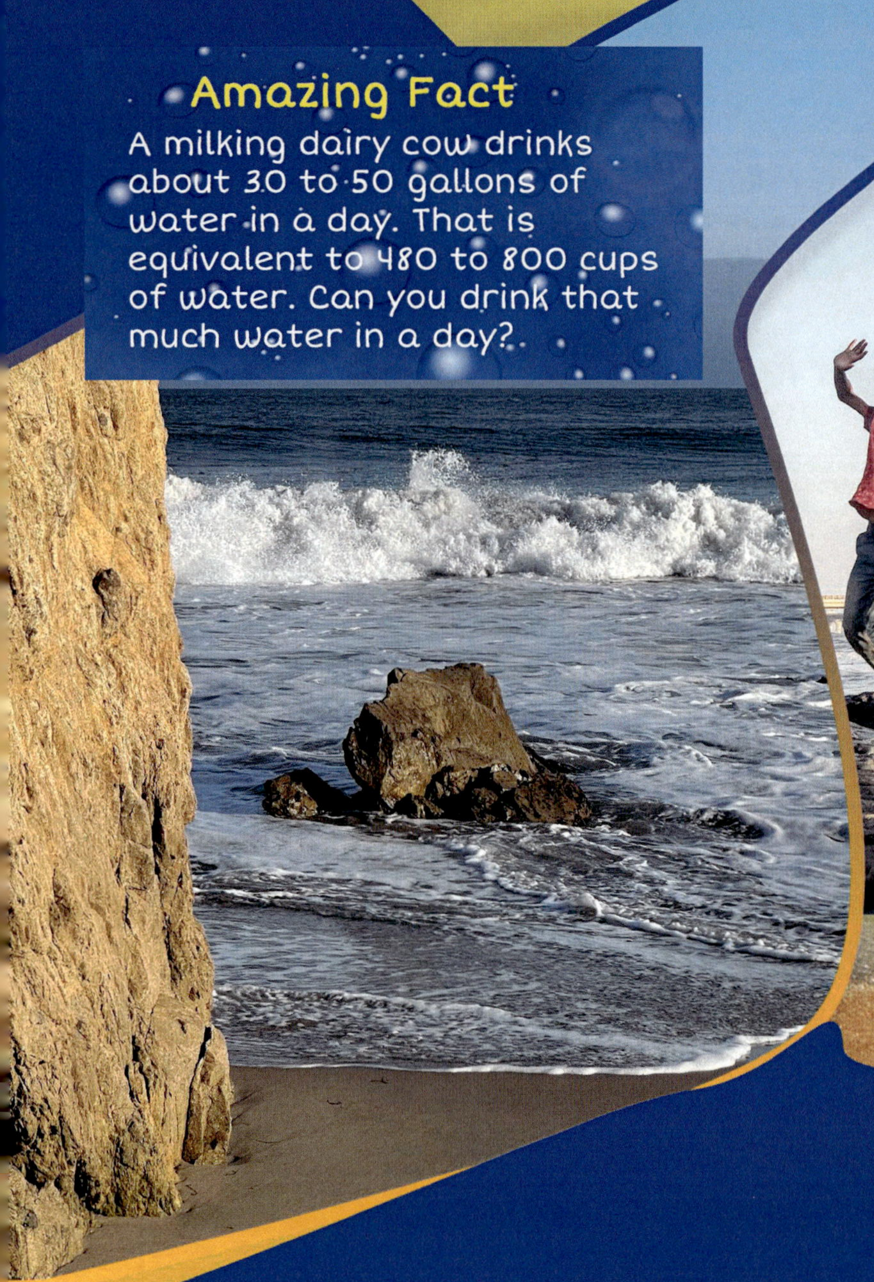

Tides originate far out in the ocean as long waves that travel across the sea. They are caused by gravitational forces exerted by the moon, sun and rotation of the earth, causing sea levels to rise and fall.

There are four types of tides:
high, low, semidiurnal and diurnal tides.
High tides occur where the sea is at the highest level.
When the tides rise, water flows toward the shore.

This is called a flood current.
Low tides occur where the sea is at its lowest level.
When the tides recede, water moves away from the shore.
This is called an ebb current.

Amazing Fact

The Amazon River in South America is the largest in the world, and home to the largest snakes on earth known as the Anaconda.

Semidiurnal tides happen when two high and two low tides occur in a day.
Diurnal tides happen when only one high and one low tide occur in a day.
This cycle happens over and over again throughout the day like a steady fountain to form a buoyant tidal cycle.

Review Questions

1. Tides are formed by gravitational forces exerted on the earth by the moon, sun and the rotation of the earth. *True or False.*

2. Diurnal and low tides are two types of tides. *True or False.*

3. When water recedes from the shore back into the ocean it is known as ______________________.

Water Habitat

The bodies of water on planet Earth are home
to many creatures and plants
that bend, creep, and drink from water's throne.
We call the places where they live their habitat.
A whole lot more species in lakes, ponds, rivers and seas like
sharks, oysters, octopuses, lobsters, hippopotamuses,
whales, seals, dolphins, eels, sea turtles and walruses
are among species which cannot live anywhere else but where
precious water drips or swims with cheer.
The water lily and floating heart are just a few plants
that can only survive where water leads.

Review Questions

1. Name two creatures which must live in water in order to survive.
   a. ________________ b. ________________

2. Water lily and the floating heart are plants that live on or around water. *True* or *False*.

3. What is a habitat?

Amazing Fact

It takes about 1,000 tons of water to produce one ton of rice which is one of the main foods eaten in Asia and used to make delicious dishes.

Water states

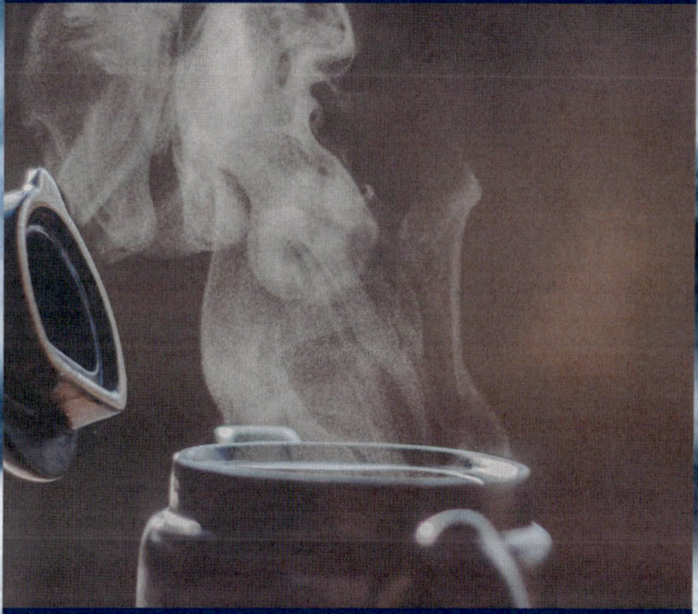

Water exists in all three states of matter
as solid, liquid and gas.
Though different the states may be,
one thing will always stay the same,
that water as solid, liquid, and vapor
is made up of the same number and kinds of atoms.

Amazing Fact

A single drop of water contains billions of water molecules.

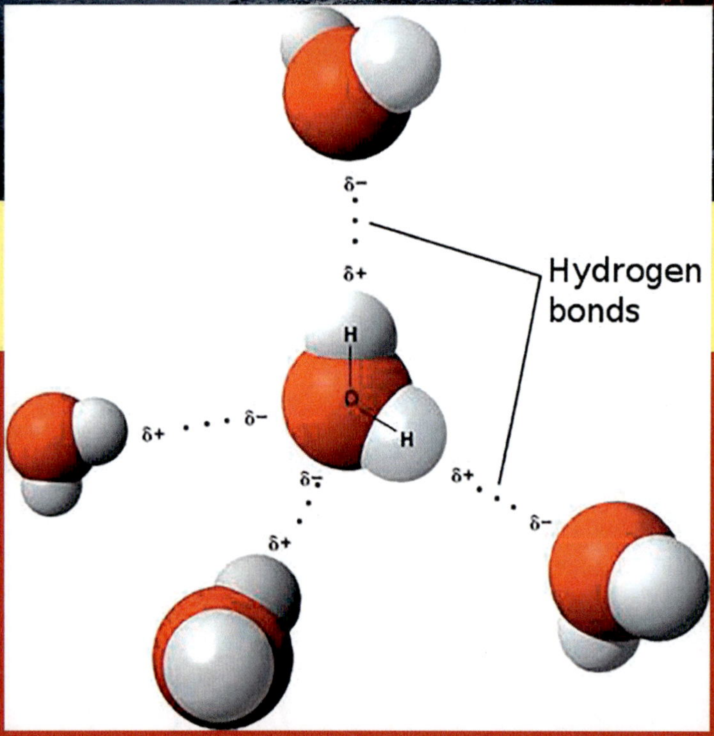

One atom of oxygen and two atoms of hydrogen bond to form one molecule of water written as H_2O. The oxygen atom has a slightly negative charge while the hydrogen atoms are slightly positive. And since opposite charges always attract, that's exactly what they set out to do.

As molecules of water interact, the oxygen and hydrogen of one molecule are attracted to atoms of nearby molecules with opposite charges, and are held together by special intermolecular forces of attraction which we call hydrogen bonds. Intermolecular forces are attractions and repulsions between atoms of molecules situated closely together.

Amazing Fact

The water in the oceans make up 97 percent of the water on earth, and the water in them is salty.

In the solid state, water exists as cubes hard as ice whose molecules are tightly packed, and are held together by strong intermolecular forces. This tight packing gives ice cubes their fixed shapes. But when heated solid ice can lose its shape. And when at 32°F [0°C], it melts into liquid water.

In the liquid state, molecules are loosely linked together and move about freely. Their intermolecular forces are not as strong as those found in solid ice. This is why water flows about easily and does not have a definite shape. Instead, it takes on the shape of any container it is in.

Amazing Fact

Water expands when it freezes, unlike most liquids which contract.

Water in its gaseous state can be seen as steam or vapor, where molecules are separated and held together by weak intermolecular forces which make it easy for the molecules to move about more freely without any hinder. When water is heated to its boiling point at 212°F [100°C], it doesn't hesitate to change into vapor.

Amazing Fact

If every household in the United States had a faucet that dripped once every second, 928 million gallons of water would go to waste every day.

Review Questions

1. The molecules in water are held together by __________ bonds.
   (a) Oxygen (b) Hydrogen (c) Carbon

2. When the temperature of water reaches 100 degrees Celsius, it turns into __________ (a) Solid (b) Liquid (c) Vapor

3. What are the atoms that make up water?

Water Types

In nature, water exists in two forms as fresh and salt water. Saltwater is the most abundant on earth, found in oceans and special lakes while freshwater is found in creeks, springs, rivers, some lakes and frozen glaciers.

Some creatures can only in saltwater survive
while others can only in freshwater stay alive.
People can drink from freshwater sources
but cannot do so from saltwater sources.

Review Questions

1. Fresh and saltwater are the only forms in which water exists. *True* or *False.*
2. Some sea creatures can survive in saltwater but not in fresh water. *True* or *False.*
3. What is the most abundant kind of water on earth? (a) Freshwater (b) Saltwater (c) Water Vapor

Amazing Fact

Water makes up 70 percent of the human body. Our blood is made up of 83 percent water, our brain 70 percent, and our lungs 83 percent.

Water Uses

Water is an essential natural resource that all living things, plants and animals, need to live. Unlike sea creatures, humans don't live in water. Yet, they cannot live without it either.

Amazing Fact

A person needs to drink about half their body weight, in ounces, of water daily.

Water serves many functions.
We swim and play in it,
and sail our ships on it.
We farm and bathe with it,

With water we wash our hands to get rid of germs,
and water our plants, grass and flowers and ferns.
We use water to brush our teeth to keep them from decay,
and also use it to clean our toys before we put them away.

We drink water to quench our thirst,
and for our parents to cook our food and put nutrition first
so we can eat, and run and grow.
There's so much wonder about water to know.

Water brings so much peace when on the beach
we watch as waves lap against our prickly feet.
The seas need lots of water to form giant waves and even the dangerous tide,
and to keep crocodiles and alligators less angry wherever they abide.

Review Questions

1. Why is water important for our planet?
2. What are some of the ways we use water?
3. People need water to live in. *True* or *false*.

Amazing Fact

A tomato is made up of about 95% water. But an apple, a pineapple, and an ear of corn each contain about 80% water.

Water Conservation

Freshwater is precious but limited in supply.
So, there's a great need to keep it going for others to have in the future.
It is the only kind of water we can drink and wash our clothes in.
So, let's not flush things we can throw away in a bin.

We use it to cook and bathe and bake.
Water indeed is a precious resource.
So, we must do our best to take care of it.
And do all we can to show we appreciate it:

The simple things we do can save tons of water and prevent waste.

While we brush our teeth and daddy shaves his beard, we can turn off the faucet and not let the water run wild. We can report leaky faucets to our parents and teachers. And turn off the faucet while washing our hair.

Amazing Fact

A person can survive for up to 3 weeks without food but only about 3-4 days without water.

Amazing Fact

Freshwater makes up approximately 3 percent of the earth's water supply but only 0.3 percent is available for drinking water. .

We can shorten the time we spend in the shower from
20 to 10 minutes or less and not near an hour.
We can dispose of dead insects in the trash or outside.
Why use precious water to flush them in a massive tide.

We can keep our rivers and oceans clean,
whether or not we are seen
by not throwing garbage in them,
and picking up trash we find on the beach.

We must do our best not to pollute them
as harmful substances create a long-term problem.
We can purify dirty water and reuse it again.
Recycling saves water and aids conservation.

Review Questions

1. Identify three ways we can conserve water.

   a. ______________________________

   b. ______________________________

   c. ______________________________

2. What are two ways we can keep our rivers clean?

   a. ______________________________

   b. ______________________________

3. Freshwater is the most abundant form of water on earth. *True* or *False.*

Water Fun Activities

Children should be supervised when performing these activities.

GAME

Water Balloon Towel Toss Game

This is a fun game for the whole family which can be done from the comfort of your backyard that is sure to keep everyone cool and giggling on a hot day.

What You'll Need:

1. Beach Towels
2. Balloons
3. Water
4. Bucket to hold the balloons
5. Hose

Directions

1. Fill up the water balloons. Divide up into two teams. Each person holds on to the end corners of the towel.
2. Place the filled balloons in the center of the towel.
3. Toss the balloon high into the air toward the other team. The other team has to try to catch it without getting soaked. If the balloon survives without breaking, they can toss it back to the first team. The team who breaks the least number of balloons wins.

ART

Ice Cube Painting

This is an all-inclusive interactive activity where kids not only get to create their own masterpieces out of water but also get firsthand experience observing a physical change reaction. This is also a good time to discuss the different states of water.

What You'll Need:

1. Blue, orange, green and red food coloring
2. 1 cup of water
3. Ice tray
4. Popsicle sticks
5. Plain white paper
6. Old newspaper

Directions

1. Pour water into the tray cube spaces evenly.
2. Add a drop of blue food coloring to four different cube spaces and mix well.
3. To another set of four different cube spaces, add a drop of blue food coloring and mix well.
4. Do the same thing for the remaining colors.
5. Place a popsicle stick into the cubes with the colored water.
6. Freeze for about 4 to 6 hours.
7. Place a plain white paper over the newspaper to prevent the surface of the table or counter from getting stained.

EXPERIMENT

Making Rain Cloud Experiment

This is a fun way to teach your child about the weather and how rain forms.

Making Rain Cloud Experiment

1. Clear cup
2. Shaving Cream
3. Food coloring
4. Water

Directions

1. Fill the clear cup ¾ up with water.
2. Add a layer of shaving cream to the top of the cup to act like cloud. Do not mix.
3. Add several drops of food coloring to the top of the shaving cream.

The drops fall through the shaving cream. As the shaving scream gets very heavy, the food coloring begins to drop down into the water.

Explain to the child that when the water droplets grow heavy in the sky, gravity pulls them down as raindrops in a similar manner as what was just observed in the experiment.

Buoyancy and Density Experiment

This is a simple experiment to teach your kids about buoyancy as they make predictions about which objects, they think will sink and which ones they think will float and why.

What You'll Need:

1. Tall Jar
2. Lemons
3. Other objects to test
4. Water

Directions

1. Fill jar ¾ up with water.
2. Ask your child to guess or predict what they think will happen when you drop the unpeeled lemon in the jar.
3. Drop the lemon with the rind on in the jar.

You should notice it floats.

4. Get another lemon. Peel off the skin. Ask the child to predict what they think will happen when it is dropped in the jar. Will it sink or float?
5. Drop the peeled lemon in the jar. It sinks.

** On a few occasions, it might float. There are some reasons for that which is explained below.
Now, you can ask them why the unpeeled lemon floated but the one without the rind sank.

Explanation

The skin of the lemon is filled with small air pockets. The pockets of air weigh less than the water it displaces, causing it to float like a life jacket. When the lemon is peeled, the cracks between the lemon segments fill with water, making it denser than the water it displaces so this causes it to sink. Sometimes the peeled lemon will float either because the segments trap air in them or the pith was not completely removed which would prevent water from entering the segments.

Now, what happens if you drop a rock in water? It sinks. This is because the downward force of gravity acting on the stone is greater than the upward force of the water. If the upward force of the water is greater than the downward force of the object, the object will float.

GLOSSARY

arctic ocean The coldest, smallest and most shallow of the oceans.

atlantic ocean The second largest ocean in the world.

atom The smallest unit of matter.

bond A strong force of attraction that holds together atoms in a molecule or crystal.

condensation The process by which water vapor changes into liquid water.

conservation The prevention of the waste of a resource.

diurnal tide Occurs when only one high and low tide happen during the day.

domain A region distinctively marked by some physical feature like water and land.

ebb current The movement of tidal current of water away from the shore.

evaporation When water changes from its liquid form into vapor.

flood current The movement of the tidal current of water back to the shore.

freshwater Water that has negligible amount of salt that cannot be tasted.

gravitational force Force that attracts or brings together two objects with mass.

habitat The home where an animal or plant lives naturally.

high tide The point in the tidal cycle when the sea level is the highest.

hydrogen One of the atoms that make up water.

intermolecular forces Force of attraction or repulsion between atoms when molecules interact.

low tide The state of the tide when sea level is at its lowest.

matter Anything that has mass and takes up space.

GLOSSARY

molecule Smallest unit of a substance that has all of the physical and chemical properties of the substance.

natural resource Things found in nature that are used to support life and meet people's needs.

oxygen One of the atoms found in water.

pacific ocean The largest ocean in the world.

planet A celestial body that orbits around a star.

pollute To contaminate with harmful substances.

precipitation Water that falls to the earth as rain, hail, sleet and snow.

purify To remove contaminants from something.

recycling Convert waste into reusable material.

salt water Water that naturally contains a high level of salt as in the oceans.

semidiurnal tide Occurs where two high and two low tides occur in a day.

southern ocean The fourth largest ocean also known as the Antarctica Ocean.

tide The alternate rising and falling of the sea.

transpiration The process whereby plants release water vapor to the air.

INDEX

A
Arctic ocean *6*
Atlantic ocean *6*
Atom(s) *14, 15*

B
Bond(s) *27, 28*

C
Conservation *42, 47*

D
Diurnal tide(s) *20, 22*
Domain *4*

E
Ebb current *21*

F
Flood current *21*
Freshwater *33, 34, 42*

G
Gravitational forces *19*

H
Habitat *24*
High tide *20*
Hydrogen *27, 28*

I
Intermolecular forces *28, 29, 30*

L
Low tide(s) *21, 22*

M
Molecule *27, 28, 29, 30, 31*

N
Natural resource *36*

O
Oxygen *27, 28*

P
Pacific ocean *6*
Planet *1, 24*
Pollute *47*
Purify *47*

R
Recycling *47*

S
Salt water *33*
Semidiurnal tide *22*
Southern ocean *6*

T
Tide(s) *17, 19, 20, 21, 22, 40, 45*

PHOTO CREDIT

1. Rolling Hills by Marisa04 from Pixabay - page 2.
2. Person in the rain photo by Alicja from Pixabay - page 11.
3. Geyser by LalouBLue from Pixabay on - page 5.
4. Kids playing in water by Peter Idowu on Unsplash - page 8.
5. Kid brushing teeth photo by Shalev Cohen on Unsplash - page 36.
6. Waterfall by Jonathan Meyer on Pexels - dedication page.
7. Elephant in the bush by Mohammed Nuzrath on Pixabay - page 13.
8. Giraffe by Ramalholore on Pixabay - page 2.
9. Girl with flower by Jill Wellington - page 1.
10. Sheltering from the rain with banana leaf by Sasin Tipchai - page 4.
11. River Nile by RonPorter on Pixabay - page 7.
12. Ice cube by MurlocCra4ler on Pixabay - page 24.
13. Dolphin swimming in water by Astrid Schmid on Pixabay - page 33.
14. Shark in water by PIRO4D on Pixabay - page 33.
15. Toys in bowl by Monikal on Pixabay - page 40.
16. Cooking corn by Karolina Grabowska on Pixabay - page 41.
17. Boy drinking water by Chaucharanje on Pexels - page 39.
18. Grasshopper climbing tree by Pixabay on Pexels - page 2.
19. Boys jumping on rock by Guduru Ajay Bhargar - page 17.
20. Boy drinking water from the tap by Chauchar Anje on Pexels - page 3
21. Waste bin on Unsplash - page 43.
22. Duck near sea by Ariann on Pexels - page 34.
23. Sea anemone by Pascal Renet - page 32,
24. Plastic on river by Brian Yurasits on unsplash - page 44.
25. Faucet by Dominika Roseclay on Pexels - page 42.
26. Clothes wringing by Teona Swift on Pexels - page 40.
27. Chrocodile by Rachel Claire on Pexels - page 38.
28. Peacock by Vadim on Pexels - page 3.

PHOTO CREDIT

29. Kid running from a rocky shore by PNW Production on Pexels- page 10.
30. Kids playing by Kindel Media on Pexels - page 8.
31. Young girl sledding in the snow by Yan Krukov on Pexels - page 12.
32. Kid bathing in a bath by Kindel Media on Pexels - page 43.
33. Kids watering plants by Rodnae Productions on Pexels - page 36.
34. Turtle by Daniel torobekov on Pexels - page 32.
35. Jelly Fish by Guillaume Meurice on Pexels - page 32.
36. Water Molecule Photo - page 25
    License--https://creativecommons.org/licenses/by-sa/3.0/
37. Water Cycle by NASA - page 9
    License--https://creativecommons.org/licenses/by-sa/3.0/

Made in the USA
Monee, IL
12 May 2025

17281804R00036